Yohaku Book 2

Mike Jacobs

Copyright © 2019 Mike Jacobs

All rights reserved.

ISBN-13: 9781093415209

For:

Màiri and Andrew;

and

my brothers and sisters: Paul, Philip, Stephen, Margaret and Elisabeth. Your brilliance inspires me and fills me with sibling pride;

and

my mother, who really was my first maths teacher;

and

my dear, late father…*cuiridh mi clach air do chàrn.*

CONTENTS

	Acknowledgments	i
1	What is a yohaku?	p. 1
2	How to solve a yohaku	p. 3
3	Puzzles	p. 6
4	Solutions	p. 67

ACKNOWLEDGMENTS

Thanks must go out to all those who have encouraged me to write this book including my wife, Jackie, and my wonderful children Màiri and Andrew without whom…

I must also acknowledge all the wonderful maths colleagues that I have had the privilege of working with over the years in Yorkshire and in Ontario: you know who you are!

Finally, thank you to all the kind people who follow my Twitter account (@YohakuPuzzle) for all their kind words of support and encouragement over the past few years. In particular, I am grateful to all the dedicated teachers who have shared with me all the different ways that they have used yohaku puzzles with their students: you truly are inspirational.

1 WHAT IS A YOHAKU?

Yohaku™ is a new type of number puzzle that will test your number sense and problem solving skills. Each yohaku is either an additive or a multiplicative puzzle (as indicated by the symbol in the bottom right of the grid). Your task is to fill in the empty cells such that they give the sum or product shown in each row and column at the same time as satisfying any restrictions that are stated below each puzzle.

Yohakus are great for use in classrooms as a lesson starter, or for the morning commute, or for something to accompany an afternoon cup of tea. Some yohakus will take a few minutes to solve, others will have you puzzling for longer.
Each yohaku is lovingly created by hand!

I'm often asked why I named these puzzles *yohaku*. A few years ago, I tried to create a number puzzle and I was told by a friend to 'Make sure it has a catchy name.' This stumped me for a while but that day I was listening to a cricket game between Yorkshire (where I was born) and Hampshire (my favourite team). I absent-mindedly jotted down the first two letters from both teams: Y-O-H-A. Suddenly I thought how these puzzles required knowledge and understanding so I put a K and a U on the end: YOHAKU! I showed this to my friend who told me that it sounded like a proper puzzle, like Sudoku!

I wasn't happy with that particular puzzle though and

put it to one side. Later I got the idea for a new puzzle when playing around with some numbers and so re-used the name *yohaku*. I first tested these puzzles on my own children. When they told me that the name sounded Japanese, it got me thinking: what if yohaku *is* a Japanese word? More importantly, what if it is a bad Japanese word?! I quickly looked it up.

It turns out that yohaku means 'blank space'. I can't think of a better word to describe these puzzles!

2 SOLVING A YOHAKU

To solve a yohaku, you must fill in the blank space so that the cells give the sum or the product shown in each row and column.
In addition, each yohaku has a further restriction.

For example, in this multiplicative yohaku (shown right) the restriction states 'Use only whole numbers'. You might look at the top row and start thinking of two numbers that multiply to give 40. Initially you might try 5 and 8.

5	8	40
		96
32	120	✗

Use only whole numbers

However, now you must think of a number that you multiply 5 by to give 32. You can't do this using whole numbers, so you will have to try a different pair of numbers, say 4 and 10. This does allow you to use whole numbers to complete

4	10	40
8		120
32	150	✗

Use only whole numbers

the problems (as hinted at above). This particular yohaku also has a second solution. Can you find it?

Generally, looking carefully at the sum or product in each row or column will give you some clues as to how these values can be decomposed and, combined with information from other rows and columns, will help you find the values for certain cells. For example, in this 3-by-3 yohaku, the restriction is 'Use prime numbers'. Looking at the top row, you need to think of three prime numbers that multiply to give 8.

There is only one way to do this i.e. 2×2×2. Putting these in, we can now look at the left-hand column for further clues: what values could we use so that 2 times a prime times a prime gives 50?

Again, there is only one way of doing this as shown on the right. With a little more thought, it should be possible to fill in the remaining cells.

Yohaku Book 2

Some yohakus will have more than one possible solution. Many will involve a fair amount of trial and error.
The puzzles in this book are arranged from the easiest to the hardest. Solutions are provided. Happy solving!

Some Definitions

Consecutive numbers: Integers from a sequence created by counting by ones
e.g. {3, 4, 5, 6}, {-3, -2, -1, 0, 1, 2, 3, 4, 5}

Prime numbers: Whole numbers with exactly two factors
e.g. 2, 3, 5, 7, 11, 13, 17, 19, 23, 29, 31, 37, 41, 43, ...

Square numbers: Whole numbers created by multiplying a whole number by itself
e.g. 0, 1, 4, 9, 16, 25, 36, 49, 64, 81, 100, 121, 144, ...

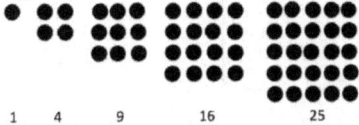

Triangle numbers: Whole numbers created by arranging dots in an equilateral triangle
e.g. 0, 1, 3, 6, 10, 15, 21, 28, 36, 45, 55, 66, 78, 91, ...

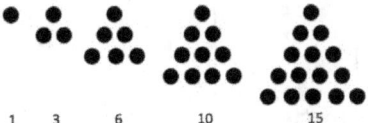

3 PUZZLES

1

		47
		51
48	50	+

Use 4 consecutive numbers

2

		121
		117
120	118	+

Use 4 consecutive numbers

3

		197
		193
194	196	+

Use 4 consecutive numbers

4

		-1
		3
0	2	+

Use 4 consecutive numbers

5

		70
		66
64	72	+

Use 4 consecutive odd numbers

6

		220
		212
214	218	+

Use 4 consecutive odd numbers

7

		100
		96
94	102	+

Use 4 consecutive even numbers

8

		396
		400
394	402	+

Use 4 consecutive even numbers

9

		10
		42
38	14	+

Use 4 different prime numbers

10

		12
		28
22	18	+

Use 4 different prime numbers

11

		22
		26
28	20	+

Use 4 different prime numbers

12

		30
		36
38	28	+

Use 4 different prime numbers

13

		54
		13
34	33	+

Use 4 different prime numbers

14

		90
		55
45	100	+

Use 4 different prime numbers

15

		25
		130
90	65	+

Use 4 different square numbers

16

		65
		74
89	50	+

Use 4 different square numbers

17

		125
		90
85	130	+

Use 4 different square numbers

18

		185
		85
125	145	+

Use 4 different square numbers

19

		269
		521
500	290	+

Use 4 different square numbers

20

		169
		101
145	125	+

Use 4 different square numbers

21

		24
		35
30	28	✗

Use 4 consecutive numbers

22

		88
		90
99	80	✗

Use 4 consecutive numbers

23

		30
		27
45	18	✗

Sum of 4 cells is 23

24

		30
		27
45	18	✗

Sum of 4 cells is 29

25

		48
		36
24	72	✗

Sum of 4 cells is 29

26

		48
		36
24	72	✗

Sum of 4 cells is 63

27

		48
		60
80	36	✗

Sum of 4 cells is 36

28

		75
		144
100	108	✗

Sum of 4 cells is 68.

29

		144
		56
84	96	✗

Sum of 4 cells is 39.

30

		36
		65
20	117	✗

Sum of 4 cells is 31.

31

		24
		-15
-30	12	✗

Sum of 4 cells is 8.

32

		24
		-28
21	-32	✗

Sum of 4 cells is 14.

33

		-24
		8
-12	16	✗

Sum of 4 cells is -1.

34

		-18
		-20
10	36	✗

Sum of 4 cells is 6.

35

		21
		-26
14	-39	✗

Sum of 4 cells is 1.

36

		24
		-24
12	-48	✗

Sum of 4 cells is 0.

37

		3
		10
5	6	✕

Sum of 4 cells is 10.

38

		14
		15
10	21	✕

Sum of 4 cells is 16.

39

		36
		3
4	27	✗

Sum of 4 cells is 19.

40

		3
		2.1
7	0.9	✗

Sum of 4 cells is 14.

41

			9
			15
			21
21	17	7	+

Use 9 consecutive numbers

42

			19
			14
			30
23	16	24	+

Use 9 consecutive numbers

43

			19
			28
			25
31	15	26	+

Use 9 consecutive numbers

44

			23
			26
			32
36	21	24	+

Use 9 consecutive numbers

45

			40
			44
			33
46	31	40	+

Use 9 consecutive numbers

46

			47
			37
			42
38	37	51	+

Use 9 consecutive numbers

47

			29
			34
			45
35	41	32	+

Use 9 consecutive numbers

48

			55
			46
			52
44	49	60	+

Use 9 consecutive numbers

49

			20
			12
			22
25	10	19	**+**

Use 9 consecutive numbers

50

			55
			40
			49
50	42	52	**+**

Use 9 consecutive numbers

51

			72
			42
			120
84	16	270	✗

Use 9 consecutive numbers.

52

			8
			144
			315
160	108	21	✗

Use 9 consecutive numbers.

53

			112
			12
			270
288	42	30	✗

Use 9 consecutive numbers.

54

			84
			320
			135
280	24	540	✗

Use 9 consecutive numbers.

55

			360
			240
			42
135	160	168	✕

Use 9 consecutive numbers.

56

			560
			165
			216
990	84	240	✕

Use 9 consecutive numbers.

57

			231
			270
			320
72	280	990	✗

Use 9 consecutive numbers.

58

			360
			420
			132
240	252	330	✗

Use 9 consecutive numbers.

59

			0
			120
			168
0	70	48	✗

Use 9 consecutive numbers.

60

			16
			140
			0
60	12	0	✗

Use 9 consecutive numbers.

61

			15
			12
			0
4	9	14	+

Use 9 consecutive numbers

62

			1
			8
			18
9	13	5	+

Use 9 consecutive numbers

63

			13
			-2
			7
0	8	10	+

Use 9 consecutive numbers

64

			10
			-5
			4
5	-3	7	+

Use 9 consecutive numbers

65

			-20
			0
			84
30	0	84	✗

Use 9 consecutive numbers.

66

			40
			0
			-6
0	-8	-15	✗

Use 9 consecutive numbers.

67

			48
			0
			2
16	0	9	✗

Use 9 consecutive numbers.

68

			4
			-9
			0
0	30	8	✗

Use 9 consecutive numbers.

69

			22
			45
			33
53	37	10	+

Use 9 different prime numbers

70

			19
			67
			43
23	45	61	+

Use 9 different prime numbers

Yohaku Book 2

71

			63
			69
			23
71	49	35	+

Use 9 different prime numbers

72

			55
			25
			65
77	21	47	+

Use 9 different prime numbers

73

			10
			39
			113
85	57	20	+

Use 9 different prime numbers

74

			27
			42
			55
65	12	47	+

Use 9 different prime numbers

75

			198
			280
			8
96	308	15	✕

Use 9 different whole numbers.

76

			84
			27
			220
231	24	90	✕

Use 9 different whole numbers.

77

			30
			88
			189
77	108	60	✗

Use 9 different whole numbers.

78

			1056
			36
			20
144	16	330	✗

Use 9 different whole numbers.

79

			1980
			70
			12
220	36	210	✗

Use 9 different whole numbers.

80

			38
			216
			150
380	18	180	✗

Use 9 different whole numbers.

81

			19
			91
			55
79	67	19	**+**

Use 9 different triangle numbers

82

			70
			34
			115
91	24	104	**+**

Use 9 different triangle numbers

83

			200
			178
			21
110	65	224	+

Use 9 different square numbers

84

			59
			101
			161
53	155	113	+

Use 9 different square numbers

85

		$\frac{1}{10}$
		$\frac{11}{30}$
$\frac{4}{15}$	$\frac{1}{5}$	**+**

Use unit fractions only

86

		$\frac{19}{60}$
		$\frac{8}{15}$
$\frac{2}{5}$	$\frac{9}{20}$	**+**

Use unit fractions only

87

		$\frac{5}{16}$
		$\frac{11}{24}$
$\frac{19}{48}$	$\frac{3}{8}$	**+**

Use unit fractions only

88

		$\frac{1}{6}$
		$\frac{5}{12}$
$\frac{7}{36}$	$\frac{7}{18}$	**+**

Use unit fractions only

89

		1
		$\frac{14}{15}$
$1\frac{2}{5}$	$\frac{2}{3}$	✗

Sum of 4 cells is 5

90

		$\frac{1}{4}$
		$\frac{7}{8}$
$1\frac{3}{4}$	$\frac{1}{8}$	✗

Sum of 4 cells is 4

91

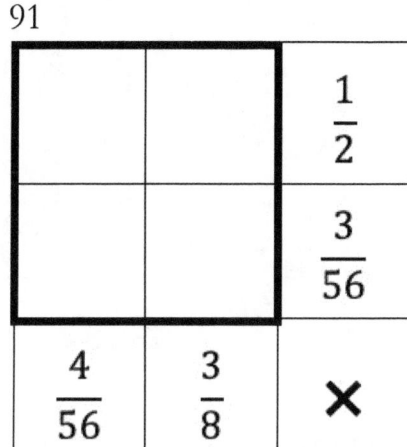

Sum of 4 cells is 2

92

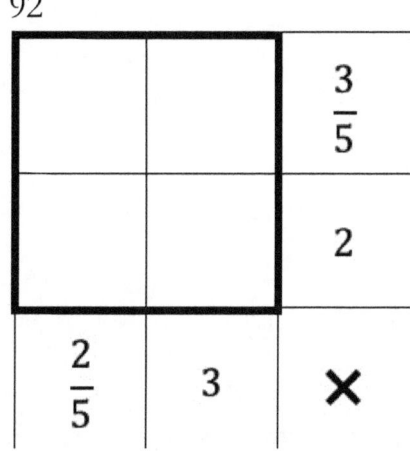

Sum of 4 cells is 5

93

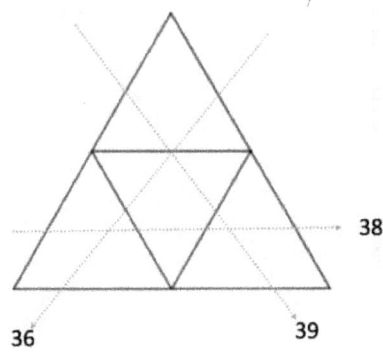

Place four consecutive numbers into the triangles to give the totals shown in each of the three rows.

94

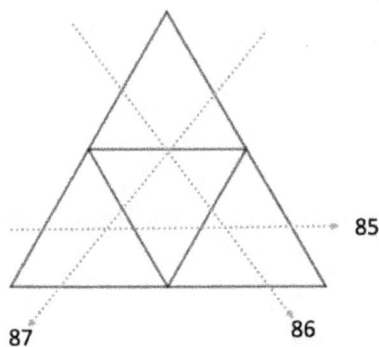

Place four consecutive numbers into the triangles to give the totals shown in each of the three rows.

95

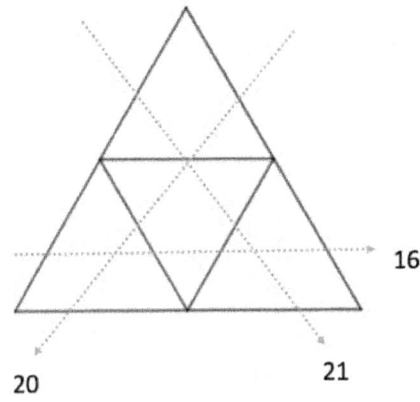

Place four different prime numbers into the triangles to give the totals shown in each of the three rows.

96

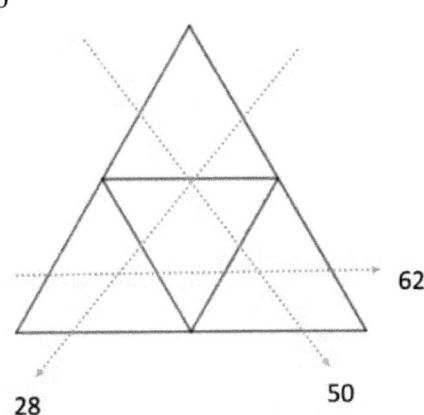

Place four different prime numbers into the triangles to give the totals shown in each of the three rows.

97

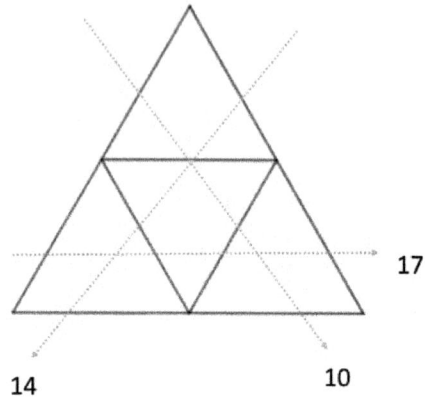

Place four different triangle numbers into the triangles to give the totals shown in each of the three rows.

98

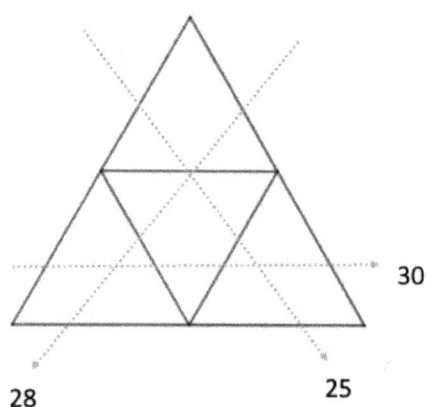

Place four different triangle numbers into the triangles to give the totals shown in each of the three rows.

99

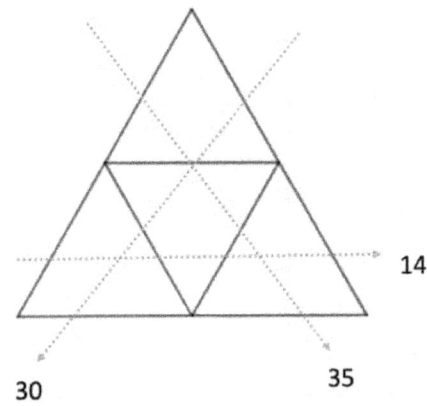

Place four different square numbers into the triangles to give the totals shown in each of the three rows.

100

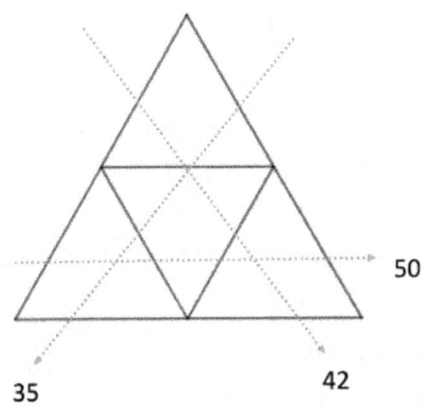

Place four different square numbers into the triangles to give the totals shown in each of the three rows.

101

				43
				45
				37
				11
18	39	29	50	+

Use 16 consecutive numbers

102

				50
				27
				35
				24
37	57	18	24	+

Use 16 consecutive numbers

103

				42
				2
				24
				36
25	11	32	36	+

Use 16 consecutive numbers

104

				-6
				13
				0
				33
2	1	11	26	+

Use 16 consecutive numbers

105

				104
				210
				770
				54
546	36	150	308	✗

Use 16 prime numbers

106

				460
				1078
				220
				260
644	484	350	260	✗

Use 16 prime numbers

107

				-6
				-624
				5040
				0
70	576	880	0	✗

Use 16 consecutive numbers

108

				160
				108
				0
				-210
210	0	-480	-14	✗

Use 16 consecutive numbers

109

				44
				43
				11
				38
30	36	20	50	+

Use 16 consecutive numbers.
Shaded cells contain four triangle numbers.

110

				10
				40
				24
				46
18	22	34	46	+

Use 16 consecutive numbers.
Shaded cells contain odd numbers.

111

				45
				41
				33
				17
54	39	28	15	+

Use 16 consecutive numbers.
Shaded cells contain square numbers

112

				42
				30
				14
				50
22	24	40	50	+

Use 16 consecutive numbers.
Shaded cells contain even numbers

113

				46
				40
				6
				28
36	17	36	31	+

Use 16 consecutive numbers.
Shaded cells contain prime numbers

114

				84
				58
				74
				48
78	76	60	50	+

Use 16 consecutive numbers.
Shaded cells contain odd numbers

115

				52
				52
				35
				29
18	42	53	55	+

Use 16 consecutive numbers.
Shaded cells contain triangle numbers

116

				102
				112
				86
				76
106	88	104	78	+

Use 16 consecutive numbers.
Shaded cells contain even numbers

117

				8
				10
				19
				3
4	33	10	-7	+

Use 16 consecutive numbers.
Shaded cells contain triangle numbers

118

				1
				-14
				-6
				11
1	-15	-16	22	+

Use 16 consecutive numbers.
Shaded cells contain prime numbers

119

					68
					11
					59
					87
					75
77	30	74	52	67	**+**

Use 25 consecutive numbers.
Shaded cells contain prime numbers.

120

					143
					59
					111
					147
					165
171	25	147	99	183	**+**

Use 25 consecutive odd numbers.
Shaded cells contain prime numbers.

4 SOLUTIONS

Remember that some yohakus can have more than one solution!

1

23	24	47
25	26	51
48	50	+

2

61	60	121
59	58	117
120	118	+

3

98	99	197
96	97	193
194	196	+

4

-1	0	-1
1	2	3
0	2	+

5

33	37	70
31	35	66
64	72	+

6

109	111	220
105	107	212
214	218	+

7

48	52	100
46	50	96
94	102	+

8

196	200	396
198	202	400
394	402	+

9

7	3	10
31	11	42
38	14	+

10

5	7	12
17	11	28
22	18	+

11

5	17	22
23	3	26
28	20	+

12

7	23	30
31	5	36
38	28	+

13

23	31	54
11	2	13
34	33	+

14

43	47	90
2	53	55
45	100	+

15

9	16	25
81	49	130
90	65	+

16

64	1	65
25	49	74
89	50	+

17

4	121	125
81	9	90
85	130	+

18

121	64	185
4	81	85
125	145	+

19

100	169	269
400	121	521
500	290	+

20

144	25	169
1	100	101
145	125	+

21

6	4	24
5	7	35
30	28	×

22

11	8	88
9	10	90
99	80	×

23

5	6	30
9	3	27
45	18	×

24

15	2	30
3	9	27
45	18	×

Yohaku Book 2

25

8	6	48
3	12	36
24	72	✗

26

24	2	48
1	36	36
24	72	✗

27

16	3	48
5	12	60
80	36	✗

28

25	3	75
4	36	144
100	108	✗

29

12	12	144
7	8	56
84	96	✗

30

4	9	36
5	13	65
20	117	✗

31

6	4	24
-5	3	-15
-30	12	✗

32

3	8	24
7	-4	-28
21	-32	✗

33

-12	2	-24
1	8	8
-12	16	✗

34

-2	9	-18
-5	4	-20
10	36	✗

35

-7	-3	21
-2	13	-26
14	-39	✗

36

-6	-4	24
-2	12	-24
12	-48	✗

37

2	1.5	3
2.5	4	10
5	6	×

38

4	3.5	14
2.5	6	15
10	21	×

39

8	4.5	36
0.5	6	3
4	27	×

40

10	0.3	3
0.7	3	2.1
7	0.9	×

41

5	3	1	9
7	6	2	15
9	8	4	21
21	17	7	+

42

7	4	8	19
5	3	6	14
11	9	10	30
23	16	24	+

43

8	4	7	19
12	6	10	28
11	5	9	25
31	15	26	+

44

11	5	7	23
12	6	8	26
13	10	9	32
36	21	24	+

45

16	10	14	40
17	12	15	44
13	9	11	33
46	31	40	+

46

15	14	18	47
10	11	16	37
13	12	17	42
38	37	51	+

47

9	12	8	29
11	13	10	34
15	16	14	45
35	41	32	+

48

16	18	21	55
13	14	19	46
15	17	20	52
44	49	60	+

Yohaku Book 2

49

10	3	7	20
6	2	4	12
9	5	8	22
25	10	19	+

50

20	16	19	55
13	12	15	40
17	14	18	49
50	42	52	+

51

4	2	9	72
7	1	6	42
3	8	5	120
84	16	270	×

52

4	2	1	8
8	6	3	144
5	9	7	315
160	108	21	×

53

8	7	2	112
4	1	3	12
9	6	5	270
288	42	30	×

54

7	2	6	84
8	4	10	320
5	3	9	135
280	24	540	×

55

9	10	4	360
5	8	6	240
3	2	7	42
135	160	168	×

56

10	7	8	560
11	3	5	165
9	4	6	216
990	84	240	×

57

3	7	11	231
6	5	9	270
4	8	10	320
72	280	990	×

58

8	9	5	360
10	7	6	420
3	4	11	132
240	252	330	×

59

0	2	1	0
3	5	8	120
4	7	6	168
0	70	48	×

60

2	1	8	16
5	4	7	140
6	3	0	0
60	12	0	×

61

3	5	7	15
2	4	6	12
-1	0	1	0
4	9	14	+

62

0	2	-1	1
3	4	1	8
6	7	5	18
9	13	5	+

63

2	5	6	13
-2	-1	1	-2
0	4	3	7
0	8	10	+

64

5	1	4	10
-2	-3	0	-5
2	-1	3	4
5	-3	7	+

65

5	-1	4	-20
1	0	3	0
6	2	7	84
30	0	84	×

66

2	4	5	40
0	1	3	0
-3	-2	-1	-6
0	-8	-15	×

67

-4	4	-3	48
2	0	3	0
-2	1	-1	2
16	0	9	×

68

-1	2	-2	4
3	-3	1	-9
0	-5	-4	0
0	30	8	×

69

13	7	2	22
23	19	3	45
17	11	5	33
53	37	10	+

70

3	5	11	19
13	23	31	67
7	17	19	43
23	45	61	+

71

31	19	13	63
29	23	17	69
11	7	5	23
71	49	35	+

72

29	7	19	55
17	3	5	25
31	11	23	65
77	21	47	+

73

5	3	2	10
19	13	7	39
61	41	11	113
85	57	20	+

74

13	3	11	27
23	2	17	42
29	7	19	55
65	12	47	+

75

6	11	3	198
8	7	5	280
2	4	1	8
96	308	15	×

76

7	6	2	84
3	1	9	27
11	4	5	220
231	24	90	×

77

1	6	5	30
11	2	4	88
7	9	3	189
77	108	60	×

78

12	8	11	1056
3	2	6	36
4	1	5	20
144	16	330	×

79

11	18	10	1980
5	2	7	70
4	1	3	12
220	36	210	×

80

19	1	2	38
4	6	9	216
5	3	10	150
380	18	180	×

81

15	1	3	19
36	45	10	91
28	21	6	55
79	67	19	+

82

36	6	28	70
10	3	21	34
45	15	55	115
91	24	104	+

83

100	36	64	200
9	25	144	178
1	4	16	21
110	65	224	+

84

1	49	9	59
16	81	4	101
36	25	100	161
53	155	113	+

85

$\frac{1}{15}$	$\frac{1}{30}$	$\frac{1}{10}$
$\frac{1}{5}$	$\frac{1}{6}$	$\frac{11}{30}$
$\frac{4}{15}$	$\frac{1}{5}$	**+**

86

$\frac{1}{15}$	$\frac{1}{4}$	$\frac{19}{60}$
$\frac{1}{3}$	$\frac{1}{5}$	$\frac{8}{15}$
$\frac{2}{5}$	$\frac{9}{20}$	**+**

87

$\frac{1}{16}$	$\frac{1}{4}$	$\frac{5}{16}$
$\frac{1}{3}$	$\frac{1}{8}$	$\frac{11}{24}$
$\frac{19}{48}$	$\frac{3}{8}$	**+**

88

$\frac{1}{9}$	$\frac{1}{18}$	$\frac{1}{6}$
$\frac{1}{12}$	$\frac{1}{3}$	$\frac{5}{12}$
$\frac{7}{36}$	$\frac{7}{18}$	**+**

89

$\frac{3}{5}$	$\frac{5}{3}$	1
$\frac{7}{3}$	$\frac{2}{5}$	$\frac{14}{15}$
$1\frac{2}{5}$	$\frac{2}{3}$	**×**

90

$\frac{2}{3}$	$\frac{3}{8}$	$\frac{1}{4}$
$\frac{21}{8}$	$\frac{1}{3}$	$\frac{7}{8}$
$1\frac{3}{4}$	$\frac{1}{8}$	**×**

91

$\frac{4}{7}$	$\frac{7}{8}$	$\frac{1}{2}$
$\frac{1}{8}$	$\frac{3}{7}$	$\frac{3}{56}$
$\frac{4}{56}$	$\frac{3}{8}$	**×**

92

$\frac{1}{3}$	$\frac{9}{5}$	$\frac{3}{5}$
$\frac{6}{5}$	$\frac{5}{3}$	2
$\frac{2}{5}$	3	**×**

93

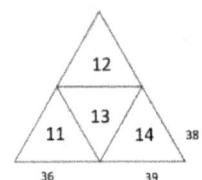

94

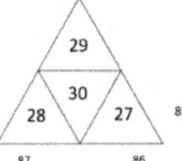

95

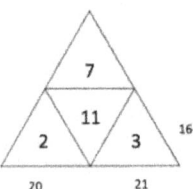

96

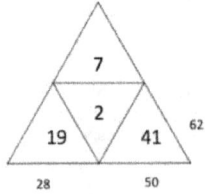

97

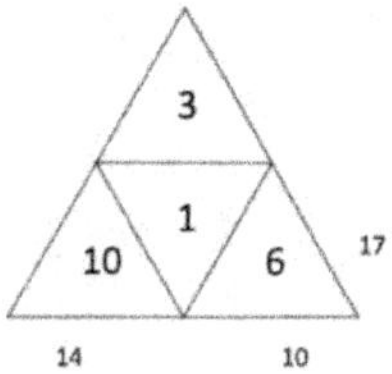

Triangle with top 3, middle row 10, 1, 6; bottom edges labeled 14, 10, right side 17.

98

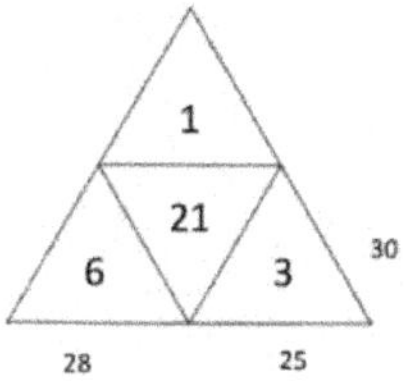

Triangle with top 1, middle row containing 21, and below 6, 3; bottom edges 28, 25, right side 30.

99

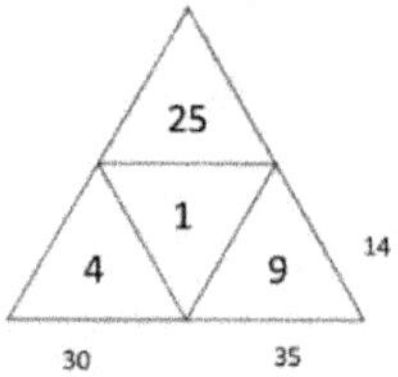

Triangle with top 25, middle 1, bottom 4, 9; bottom edges 30, 35, right side 14.

100

Triangle with top 1, middle 25, bottom 9, 16; bottom edges 35, 42, right side 50.

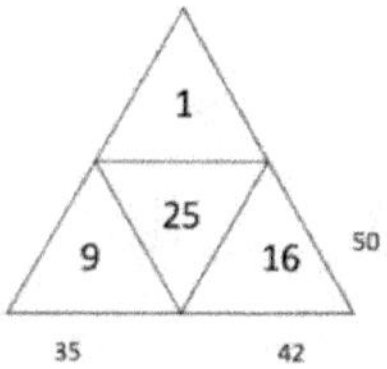

101

6	12	10	15	43
7	13	9	16	45
4	11	8	14	37
1	3	2	5	11
18	39	29	50	+

102

13	16	10	11	50
8	14	1	4	27
9	15	5	6	35
7	12	2	3	24
37	57	18	24	+

103

10	5	13	14	42
0	-1	1	2	2
6	3	7	8	24
9	4	11	12	36
25	11	32	36	+

104

-5	-4	-1	4	-6
2	1	3	7	13
-3	-2	0	5	0
8	6	9	10	33
2	1	11	26	+

105

13	2	2	2	104
2	3	5	7	210
7	2	5	11	770
3	3	3	2	54
546	36	150	308	×

106

23	2	2	5	460
7	11	7	2	1078
2	11	5	2	220
2	2	5	13	260
644	484	350	260	×

107

-1	1	2	3	-6
-2	6	4	13	-624
7	8	10	9	5040
5	12	11	0	0
70	576	880	0	×

108

-5	8	4	-1	160
-3	-6	6	1	108
-7	0	-4	2	0
-2	3	5	7	-210
210	0	-480	-14	×

109

9	13	7	15	44
11	10	6	16	43
2	1	3	5	11
8	12	4	14	38
30	36	20	50	+

110

1	0	2	7	10
5	8	12	15	40
3	4	6	11	24
9	10	14	13	46
18	22	34	46	+

111

16	13	12	4	45
15	11	8	7	41
14	10	6	3	33
9	5	2	1	17
54	39	28	15	+

112

6	7	13	16	42
4	5	9	12	30
2	1	3	8	14
10	11	15	14	50
22	24	40	50	+

113

15	6	13	12	46
14	5	11	10	40
0	2	3	1	6
7	4	9	8	28
36	17	36	31	+

114

23	24	20	17	84
19	16	12	11	58
21	22	18	13	74
15	14	10	9	48
78	76	60	50	+

115

6	14	17	15	52
5	13	16	18	52
4	8	11	12	35
3	7	9	10	29
18	42	53	55	+

116

28	25	29	20	102
30	27	31	24	112
26	19	23	18	86
22	17	21	16	76
106	88	104	78	+

117

-1	9	4	-4	8
0	10	3	-3	10
7	6	1	5	19
-2	8	2	-5	3
4	33	10	-7	+

118

2	-2	-4	5	1
-3	-8	-7	4	-14
-1	-5	-6	6	-6
3	0	1	7	11
1	-15	-16	22	+

119

22	4	14	8	20	68
1	2	5	3	0	11
12	7	17	13	10	59
18	11	23	19	16	87
24	6	15	9	21	75
77	30	74	52	67	+

120

43	1	33	21	45	143
15	3	13	11	17	59
27	5	29	19	31	111
39	7	37	23	41	147
47	9	35	25	49	165
171	25	147	99	183	+

ABOUT THE AUTHOR

Mike Jacobs is a Mathematics teacher living in Ontario, Canada. Originally from Yorkshire in the U.K., he has been teaching for 29 years. He first fell in love with mathematics after watching Disney's Donald in Mathemagic Land. He created yohaku puzzles as a means of encouraging problem solving as well as practising and improving number sense. His favourite prime number is 23 and he is convinced that triangle numbers are much more cooler than square numbers.

www.ingramcontent.com/pod-product-compliance
Lightning Source LLC
Chambersburg PA
CBHW072202170526
45158CB00004BB/1742